GW01162343

British Light Cavalry

British Light Cavalry

by
John Pimlott
illustrated by
Emir Bukhari

Almark Publishing Co. Ltd., London

© 1977 Almark Publishing Co. Ltd

Text © John Pimlott

Illustrations © Emir Bukhari

All rights reserved. No part of this publication may be reproduced, stored in a retrieval system or transmitted by any means electronic, mechanical, or by photo-copying, without prior permission of the publishers.

First Published 1977.

ISBN 0 85524 271 X

Distributed in the U.S.A. by
Squadron/Signal Publications Inc.,
3515, East Ten Mile Road,
Warren, Michigan 48091.

Printed in Great Britain by
Staples Printers Ltd,
Trafalgar Road, Kettering,
Northamptonshire,
for the Publishers, Almark Publishing Co. Ltd.,
49 Malden Way, New Malden,
Surrey KT3 6EA, England.

Contents

Light Cavalry Tactics 1759-1808	7
Problems in the Peninsula	13
Tactics 1808-1815 The Charge	17
Tactics Pursuit and Retreat	23
Tactics Picquets, Reconnaissance and Skirmishing	33
Conclusion and Bibliography	45

Preface

Tactics have been defined as "the exercises and manoeuvres of an army, or corps, or detachment when engaged with the enemy". In most cases, particularly with European armies these are relatively easy to explain, for it has long been a self-appointed task of military theoreticians to discuss them in great detail, and usual for the armies involved to follow certain basic principles.

With the British army of the eighteenth and early-nineteenth century however, there is a problem. Although theoretical tactical ideas undoubtedly existed, the central authorities lacked the power to enforce their implementation within individual regiments, which were still controlled by their respective Colonels. Various attempts were made in the 1790's to impose uniformity of drill and training, some of which were moderately successful. However this did nothing to destroy a pattern of tactical development which had begun in the eighteenth and was to persist well into the twentieth century. The depressing feature of this pattern was that the tactical lessons of a particular campaign were invariably forgotten once peace returned, necessitating an often costly process of re-learning in the next war. The cost was far higher if, in addition, the regiments of a specific arm imagined themselves to be capable of a tactic for which they had not been designed and stressed that tactic above all others, despite experiences which illustrated their mistake.

The history of British light cavalry from its origins in the mid eighteenth century to the end of the Napoleonic Wars in 1815 fits into this general pattern exactly. The first permanent light dragoon units were raised in 1759 and were designed as fast moving adaptable mounted infantry, capable of reconnaissance, skirmishing and the pursuit of a broken enemy. As such they should have been extremely useful, but as the century progressed the original functions were gradually overlaid by a false and rather paradoxical emphasis upon shock action. This meant that when the Peninsula War began in 1808 light cavalry officers regarded the charge as the only tactic to be effected and totally ignored the specialist roles for which their units had been originally intended. The tactical history of such units between 1808 and 1815 is therefore one of improvisation and the forced re-learning of basic skills under arduous combat conditions. The problems encountered and the degree of success with tactics like the charge, pursuit of a defeated enemy, covering of a retreat, reconnaissance, skirmishing and outpost duty form the main themes of this book. Any conclusions which are drawn, however, have to be tempered by the fact that as soon as the French Wars ended, the old pattern reappeared. The charge was restored to its former pre-eminence and the hard won lessons of the Peninsula and Waterloo campaigns were promptly forgotten. So the circle continued.

John Pimlott
Sandhurst 1976

British Light Dragoon

Light Cavalry Tactics 1759-1808

During the eighteenth and early nineteenth centuries European armies depended as much upon manoeuvre as set piece battles for their success. In such circumstances the existence of light cavalry, capable of swift movement over difficult ground and trained for skirmishing, reconnaissance, patrolling and pursuit, was extremely advantageous. Recognition of this led to a clear distinction of cavalry functions in Britain as early as 1645, when the New Model Army mustered two types of mounted soldier. On the one hand there existed eleven regular regiments of "horse", trained and equipped for shock action in the charge, and, on the other 1,000 "dragoons", formed as mounted infantrymen, who could skirmish reconnoitre and pursue.

This distinction remained as the Royal Army developed after 1660, but with the growth of British involvement in European campaigns, the differentiation gradually blurred. At the height of the Spanish Succession War, for example, although Marlborough had at his disposal nearly 7,000 horse and 6,000 dragoons, they were numerically inferior to the corresponding units of his opponents. As a result, he tended to use them all together for the massed charge, as at Blenheim in 1704, without regard for their original functions. By 1713 any idea of British dragoon regiments acting differently from the horse had virtually disappeared.

During the long peace which followed the Treaty of Utrecht, however, the need for some form of light cavalry was reasserted. The maintenance of law and order in Britain, together with the growing need to combat smuggling, demanded fast moving, adaptable troops which the army did not possess. Dragoon regiments could not be persuaded to revert to their original functions, eschewing the excitement and glory of the charge, so new ideas had to be introduced. One of the earliest of these came from Lieutenant-General Henry Hawley who, in January 1728, presented to the Duke of Cumberland "A scheme for reviving a regiment of original Dragoons for the use they were first intended, since the modern Dragoons are become better Horse than ever was in England before". Despite its wordy title, the scheme was eminently sensible. Hawley envisaged a body of mounted infantry, equipped and trained to skirmish on horseback or on foot, mounted on cheap hardy Yorkshire nags called "bastards", and capable of extremely swift movement in areas where existing dragoons could not operate. Each trooper, of short stature and low weight, was to be issued with "a good short sword or hanger, a small saddle with one pistol on the near side, a bill or hatchet on the other, a small hunting bit, a good firelock, bayonet and sling, in a bucket". Such a regiment, it was concluded, would be especially useful in Wales, Scotland and on the coasts of England where smugglers were most active.

The scheme met with no immediate acceptance, probably because of political pressure for military economy during periods of peace, but in 1745, the Duke of Kingston raised a regiment of light cavalry, modelled upon the Hungarian hussars, specifically

7

for duty against the Jacobites in Scotland. This unit was short-lived, however, and it was not until 14 April 1756, at the beginning of the Seven Years' War, that the scheme was extended to the regular cavalry. On that date a troop of light dragoons was ordered to be added to each of the eleven regiments of horse (or, as they had been retitled in 1746, dragoon guards) and dragoons on the British establishment. These proved so advantageous that three years later seven regiments composed entirely of light cavalry were specially raised – the 15th, 16th, 17th, 18th, 19th, 20th (Inniskilling) and 21st (Royal Foresters) Light Dragoons. When peace was restored in 1763 the 17th, 20th and 21st were disbanded, with the 18th becoming the 17th and the 19th the 18th, but the restoration of light cavalry units as a permanent part of the British army had been achieved.

Contemporary military writers were unanimous in their opinion regarding the roles of these new troops. In the late 1770's both Captain Robert Hinde (formerly of the Royal Foresters), in his *Discipline of Light Horse*, and Thomas Simes, in his *Military Guide for Young Officers*, outlined "the particular Duties on which Light Cavalry are to be employed". These are worth quoting at length:

> (Light Cavalry) are to be employed in reconnoitring the enemy, and discovering his motions: and as often as Officers are detached on such commands all that will be required of them, is to make their observations with certainty, so as not to deceive the Commanding Officer afterwards by false intelligence; they are, also, on such parties, to avoid engaging with the enemy, as being sent out for a different purpose.
>
> Light cavalry are also to be made use of for distant advanced posts, to prevent the army from being falsely alarmed, or surprised by the enemy....
>
> Parties are also to be sent out to distress the enemy, by depriving them of forage and provisions; by surprising their convoys, attacking their baggage, harrassing them on their march, cutting off small detachments, and sometimes carrying off foraging parties; in short, of seizing all opportunities to do them as much damage as they possibly can.
>
> Light cavalry are moreover to be employed in raising contributions (of provisions): and, when the army marches, they may compose the advance-guard; reconnoitring the front and flanks carefully, and sending intelligence to the Commander in Chief with expedition whenever they discover the enemy, or any kind of danger; and, when other troops cannot be spared, they may form the rear-guard, or cover the baggage....

These were specialist tasks of integral importance to eighteenth century armies, and at first glance the British light dragoons would seem to have carried them out effectively. During the American War of Independence (1775-83), for example, the 16th and 17th Light Dragoons were used extensively against the rebellious colonists, acting as skirmishers, intelligence gatherers and mounted infantry in most of the major engagements. The experience thus gained was reinforced in all light cavalry units between 1783 and 1793, when they were used in both Britain and Ireland to combat smuggling and maintain law and order; duties which provided valuable lessons in detached command, reconnaissance, patrol work and swift movement over difficult terrain. Add to this that the central authorities laid down detailed regulations for the instruction of light dragoons in dismounted drill, skirmishing, firing from the saddle, field fortification and entrenchment, and there appears to be a significant development in cavalry tactics. Nor was this confined to one or two units only: by September 1783 a total of thirteen light cavalry regiments, eventually numbered 7 – 19 inclusive on the Army Lists, were in existence, giving a potential of nearly 8,000 effective troopers.

Unfortunately such a picture is incomplete. Despite detailed instructions from the Adjutant-General's Office, central control over light cavalry training was weak. Drills and tactical exercises were left to individual commanding officers to organise at regimental level, and it was a sad fact that by the 1780's and 1790's specialist functions were largely ignored, with overwhelming preference being given to the one tactic which had destroyed the effectiveness of the original dragoons – that is, the charge. This was undoubtedly a direct result of the first occasion upon which British light cavalry were committed to battle. On 16 July 1760, only three days after landing in Germany, the newly-raised 15th Light Dragoons charged directly against formed French infantry at Emsdorf, and, much to the surprise of observers, completely routed them. Once this exploit became known – and the 15th made sure that it was by emblazoning full details on their helmet-plates – other light cavalry units sought only to emulate it, relegating such duties as reconnaissance or skirmishing firmly to the background. The charge rapidly became the only tactic specifically trained for in the regiments, and the only duty to be viewed with enthusiasm by light cavalry officers. As they appeared, the lessons of 1775-83 were dismissed as inapplicable to European campaigns, while the experiences of the coast duty were regarded as mere police work of no particular tactical significance. Few officers realised the inherent dangers of the charge – that light cavalry, being lightly armed and mounted, lacked the essential weight to be decisive, depending purely upon speed and surprise which might not always be available – and few referred back to Hawley, Hinde or Simes, who had not even contemplated the rather paradoxical idea of light cavalry shock action.

Such emphasis upon the charge was reinforced at the beginning of the French Revolutionary Wars. During the Duke of York's unfortunate Flanders campaign (1793-5), an entire brigade of light cavalry saw service, but the only incident of note was a repetition of Emsdorf. On 24 April 1794, at Villers-en-Cauchies, two squadrons of the 15th Light Dragoons, together with two squadrons of Austrian Hussars, charged and dispersed six battalions of French infantry with little loss to themselves. Beyond this, none of the specialist functions for which the light dragoons had originally been intended appear to have been carried out, not even during the long retreat to Hanover in the winter of 1794-5, when flexible, fast moving mounted infantry could have been particularly useful. This state of affairs serves as a damning condemnation of light cavalry development in the late eighteenth century: indeed, if Hawley had still been alive, he would probably have described the light dragoon regiments as "better *dragoons* than ever was in England before".

The Flanders campaign was not entirely devoid of advantage, however, for while the remnants of York's army awaited transports home from Germany, two important military thinkers of the age set their minds to cavalry improvement. The first of these men, Major-General David Dundas, was already well-known in army circles for his infantry drill manual, formally introduced in 1792, which had, for the first time, imposed centrally-inspired ideas of tactical movement on all the infantry. In 1795, viewing the *débâcle* of Flanders, Dundas initiated a similar process for the cavalry, which had become notorious for its lack of uniformity in drill and tactical manoeuvre. With different units following different training schemes, it was not unknown for cavalry commanders to misunderstand crucial orders that were not expressed in their own particular regimental phraseology. This was blatantly unsatisfactory whatever the tactics employed, and Dundas took the opportunity of the enforced inaction in Hanover, where he acted as temporary Commander-in-Chief of the remains of the Flanders forces, to impose upon all mounted regiments under his control a system of drill and manoeuvre which he had previously thought out. The details were necessarily complex, for he was in effect starting from scratch, but the regulations soon proved their worth. The King subsequently directed the system to be observed by the whole of the British cavalry, and in July 1795 copies of Dundas's *Rules and Regulations* were issued by the Adjutant-General to every regiment on the home establishment.

Although this work was of the utmost importance as the first successful attempt by the central authorities to impose uniformity of drill upon the cavalry, it suffered from one major disadvantage: it applied with equal force to all mounted units, whether dragoon guard, dragoon or light dragoon. As many of these units existed specifically for, or had strong traditions in, the charge, the new regulations naturally stressed that tactic above all others, and the relatively small section which dealt with skirmishing and dismounted drill was easily ignored by light dragoon commanders. As a result Dundas unwittingly emphasised still further the trend away from light cavalry specialisation, apparently giving official sanction to the prevailing predilection for universal shock action. Soon afterwards General Sir John Money produced a pamphlet urging the need for *real* light cavalry in the British army, capable of skirmishing, reconnaissance, patrolling and pursuit, which indicates how little had really been achieved since Hawley's scheme of 1728.

In many respects the second improvement to be introduced in the 1790's further compounded the problem. Major John Gaspard Le Marchant had served with the 2nd Dragoon Guards during the Flanders campaign and had been shocked by the depressingly bad sword and horsemanship which seemed to characterise the British cavalry of the time. This was partly a result of poor training and weapon design – horses were unused to the sound of gunfire, regiments, split into small groups for the coast duty, had enjoyed few opportunities to train as coherent bodies, swords differed in design from unit to unit and had unfortunate tendencies to break off at the hilt – but it was also apparent that few cavalry troopers were natural riders or swordsmen. After some engagements with the French, for example, surgeons and veterinary officers reported alarming numbers of wounds to both men and horses which could only have been inflicted by the riders themselves in the excitement of the charge. Le Marchant noted all these deficiencies, and, while serving in the 16th Light Dragoons in 1795-6, went to great lengths to become an extremely skilled swordsman and rider himself. He then committed his experiences to paper as *Rules and Regulations for the Sword Exercises of the Cavalry*. Despite its title, this seems to have been applicable to light dragoon units only, for the exercises involved all centred around a new pattern of sabre, broad-bladed, markedly-curved and designed for cutting rather than thrusting, which had been mass-produced and issued to every light cavalry regiment, under pressure from Le Marchant, in 1796. At any event, the King approved of the exercises and in April 1797 each light cavalry officer was ordered to furnish himself with a copy of the work.

The result was a significant and speedy improvement to British light cavalry efficiency. Special schools were established in Britain and Ireland for instruction in the exercises and the new sword design proved more robust than its multifarious predecessors. Once again, however, the improvement did nothing to ensure a differentiation of function between heavy and light cavalry units. Not only were they all now trained in a uniform system of drill, but they also seemed to be preparing for the same basic tactic, for the cuts and parries contained in Le Marchant's regulations appeared to be designed specifically for use in the *mêlée* of the charge. In these circumstances, light cavalry specialisation moved even further away from the realms of probability: the charge had been unofficially favoured in the regiments since 1760, and now through the works of both Dundas and Le Marchant, it seemed to have gained official acceptance.

In fact the light cavalry units, which by 1797 numbered 7 to 25 inclusive on the Army Lists, saw little real action before 1808. Detachments were sent to the West Indies, and entire regiments saw service in such unlikely places as Corsica, Egypt, South Africa and Latin America, but the difficulties of horse-transport and the general unsuitability of terrain combined to make the various campaigns more infantry than cavalry affairs. This was unfortunate, for the period of inaction meant that the new training schedules, stressing the charge and ignoring the specialist roles, had time to become firmly established with the light cavalry. In 1806 and 1807 four light dragoon regiments – the 7th, 10th, 15th, and 18th – were officially redesignated Hussars, but no tactical change emerged: indeed, as a contemporary military dictionary points out, this merely meant that the units had been "ordered by their respective Colonels to wear moustaches, furred cloaks and caps, etc, in imitation of the Germans". Thus, when the light cavalry were shipped to the Peninsula in 1808 to begin what was to be their only sustained campaign of the French Wars, they were ill-prepared for the specialist roles which they were called upon to carry out. If they were to be at all effective, the original functions for which they were designed had to be improvised in the field: a circumstance exacerbated by a series of practical problems peculiar to the Peninsula. These problems were to shape the tactics eventually evolved, and were to affect the use of light cavalry right up to the battle of Waterloo.

Spanish Lancer La Mancha Regiment

The Iberian Peninsula

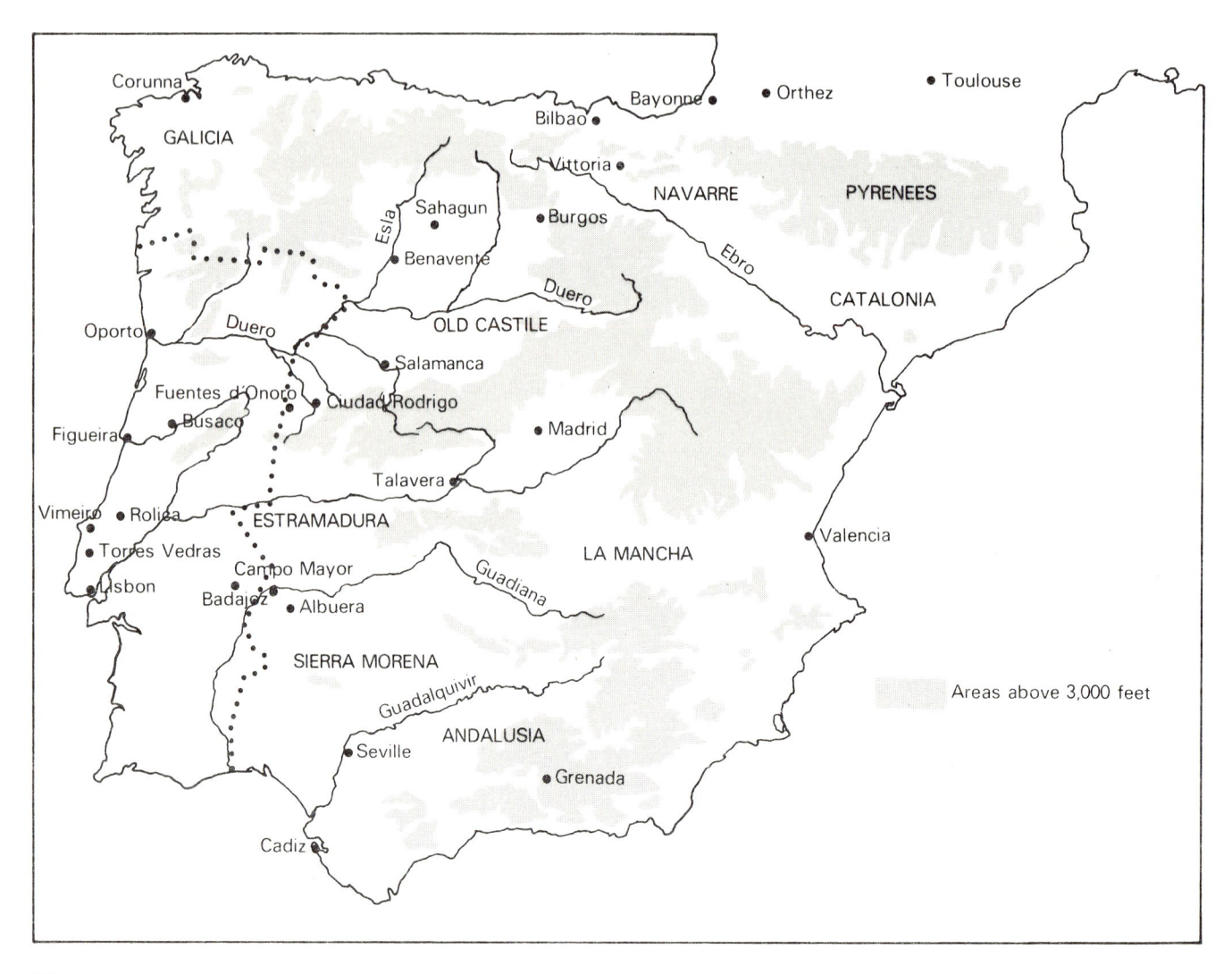

Problems in the Peninsula

Neither Sir John Moore, who commanded the British forces in the Corunna campaign of 1808-9, nor the Duke of Wellington, who was responsible for operations between 1809 and 1814, can be described as "cavalry conscious". Neither adopted a characteristic cavalry tactic on the battlefield such as that used by Napoleon, who favoured massed horsemen as a tactical reserve, available for a charge or decisive counterstroke whenever the opportunity arose, and neither displayed a great deal of enthusiasm for the mounted units under their command. This may be explained in part by the infantry orientated background and experience of both men – Moore was basically a light infantryman while Wellington, despite two years as a Lieutenant in the 12th Light Dragoons, had gained his important Indian victories mainly by infantry attack – but there was more. Throughout the Peninsula campaigns, and to a certain extent at Waterloo, a series of problems bedevilled the British light and heavy cavalry, which made their decisive use in most of the major engagements virtually impossible. As Wellington himself pointed out in 1813, "our cavalry never gained a battle yet. When the infantry have beaten the French, then the cavalry, if they can act, make the whole complete, and do wonders; but they never yet beat the French themselves".

The first and most important problem between 1808 and 1814 was the geography of the Iberian Peninsula. Vast areas of Spain – chiefly the immense Pyrennean tract, which extends from Catalonia, through Aragon and Navarre, to Asturia and Galicia – and nearly the whole of Portugal are totally unsuitable for any kind of cavalry action. High mountains, deep ravines, swift-flowing streams or rivers, and extremely difficult roads made it impossible to move mounted troops in massed formation and also endangered valuable horses. Because of such conditions, Wellington was forced to send back nearly all his cavalry to the Ebro plain during the Pyrennean campaign of 1813, while four years earlier Moore's squadrons had been virtually useless once his retreating army had entered the Galician mountains.

Even so, there were at least two areas of the Peninsula which could be termed good cavalry country – the plains between Cuidad Rodrigo and Burgos and those of Estremadura between Badajoz and the Sierra Morena, where horsemen could "ride for 20 or 30 miles without meeting any serious natural obstacle" – and it might well be imagined that in these regions the cavalry proved its worth. But for much of the time this was not the case. Before 1812-13, whenever Wellington advanced out of the mountains, as in the campaigns of Talavera (1809) and Fuentes d'Onoro (1811), he was forced to move with extreme caution and could not let his horsemen loose. The reason for this highlights the second major problem, for, with the exception of the Salamanca and Vittoria campaigns of 1812 and 1813, neither Moore nor Wellington enjoyed cavalry parity with the French. Needless to say, a superiority of horsemen was almost an unheard of luxury, confined to minor engagements of local importance.

Concentrating solely upon the light cavalry for a moment, some figures will show the extent of the problem. When Moore advanced into Spain in 1808 he had with him only two cavalry units – the 18th Hussars and 3rd Light Dragoons King's German Legion (K.G.L.). These were joined by the 7th, 10th and 15th Hussars from Britain before the retreat to Corunna began, but the force as a whole was hopelessly inferior to its French opponents. The situation was little better in 1809 when Wellington began his campaigns, for, although he did have some units of heavy horsemen under his command, he was forced to make do with four light cavalry regiments only – the 14th, 16th and 23rd Light Dragoons with the 1st Hussars K.G.L. This total remained, with the 13th replacing the 23rd after Talavera, until the advance of 1811, when the 9th, 11th and 12th Light Dragoons, together with the 2nd Hussars K.G.L., took the number up to eight, but it was not until 1813 that the light cavalry force attained what might be regarded as a reasonable strength. In that year, although the 9th and 11th Light Dragoons and the 2nd Hussars K.G.L. were sent home, their loss was more than countered by the arrival of an entire brigade, consisting of the 7th, 10th, 15th and 18th Hussars, which increased the total to nine complete units. Nonetheless, the number of available light cavalry troopers for any one campaign never exceeded three or four thousand, and this was rarely sufficient to outnumber the French.

The reason for this lack of available light horsemen is not difficult to find. One of the main arguments behind the original formation of British light dragoon regiments in the eighteenth century had been the need for adaptable, fast-moving troops on the coast duty and to aid the civil power in the home islands. When war with France began in 1793 all light dragoon units were actively involved with these duties, and, although extra regiments were raised, the importance of internal security did not diminish. Indeed, as the fear of Jacobinism in Britain grew, and the government's counter-measures became more repressive, urban rioting spread and military aid to local magistrates increased. In 1809, for example, of the nineteen regular light cavalry units on the Army Lists, a total of nine were tied down in Britain or Ireland. When it is added that a further five were stationed in India, protecting the commercial interests of the East India Company, with one unit occupying the former Dutch settlements at the Cape of Good Hope and another on the island of Sicily maintaining British influence in the Mediterranean, it is easy to see why Wellington had a mere three British regiments under his command in the Peninsula.

Even when Wellington's cavalry force was augmented in 1811 and 1813, a process made possible by the mobilisation of yeomanry and volunteer units at home which relieved the regular regiments, it soon became apparent that a lack of numbers was not the only problem to be faced. In Wellington's own words, the Peninsula was a "grave of horses". Once a regiment, either heavy or light, received orders for Peninsula service, it usually left a depot squadron in Britain as a reserve and embarked about two-thirds of its effective strength theoretically about 600 men and horses. But transportation was exceptionally poor, especially for animals. It was organised through the Transport Office at the Admiralty, which could provide no specialised vessels. Usually all that could be hired were colliers or hastily-prepared merchantmen, and it was a universal complaint among cavalry units that these were "wholly unfit for the service .... unsafe .... (and) not seaworthy". Special stalls, designed to prevent terrified horses from lashing out and breaking limbs, were not standard fitting, and as a result horse-burials at sea were depressingly frequent, even on the relatively short passage to

Lisbon or Corunna. It was not unknown for a regiment to lose a quarter of its animals even before it arrived in the Peninsula.

Once landed, the problems were by no means over. Sea sickness affected both men and horses for some days after disembarkation, necessitating a period of recuperation which precluded swift marching against the enemy, while replacements for the lost animals were impossible to find. British cavalry regiments had a reputation for being well-mounted on horses specially bred in England over centuries of mounted warfare: Peninsula animals were little more than ponies, totally unsuitable as remounts. Extra horses had therefore to be shipped in from Britain – which involved the problems of transportation all over again – or selected cavalry units, such as the 9th and 11th Light Dragoons in 1813, were sent home, leaving their mounts behind for other regiments. Neither solution was satisfactory, so great efforts had to be made within the regiments to conserve the horses they had.

Such conservation was not easy. Once a regiment started marching, two significant problems arose. The first was the provision of forage, for although certain areas of the Peninsula abounded with supplies, others did not, and the Treasury-controlled Commissariat appeared to be unable to achieve a balanced flow. Thus, if a regiment was operating in Portugal or the central plains of Spain, its horses might well receive sufficient supplies, and if they did not, their diet could be supplemented by foraging. If, on the other hand, the regiment was in the mountains or was advancing through areas already picked clean by the French, the horses could go for days eating nothing but sparce and unsatisfactory hill grass. Being used to a balanced and consistent diet of oats and hay, British horses suffered considerably in these circumstances and swiftly lost condition. This affected the mobility of the regiments, for all marches were made at a walk. William Tomkinson, an officer in the 16th Light Dragoons, pointed out, "it is impossible where you have daily a change of forage, and some days none, to keep the horses efficient if you move faster".

The second marching problem arose whatever the speed of movement, for the execrable roads of the Peninsula destroyed the horses' shoes with frightening rapidity. At first the regimental farriers, equipped with cumbersome, mule-drawn forge carts, could cope, but as their waggons broke down and they lagged behind, the horses could not be reshod, and many went lame and had to be shot. During the Corunna retreat William Vermer, an officer in the 7th Hussars, noted that his particular regiment "embarked 640 horses and brought home 60", the vast majority having been destroyed through loss of shoes. Later in the war, portable forges, carried on the backs of two mules, were specially designed and each cavalry trooper was issued with a set of shoes and nails which he carried in saddle-pouches. However in the absence of general shoeing skill these improvements did little to rectify the situation. It is small wonder, therefore, that Wellington was loath to commit his mounted forces to battle unless he could be absolutely sure of success. He realised that "Our Cavalry is the most delicate instrument in our whole machine", and that horse losses in action, when added to the phenomenal attrition rate from these other causes, could well destroy it completely in a very short time.

Such a lack of interest in cavalry usage by the army commander naturally transmitted itself to other stratas of the command structure, and, although it has been customary to describe many of Wellington's cavalry generals as nonentities, forced upon him by political pressure at home, it could well be argued that he saw little profit in employing men of worth for a portion of the service he distrusted. Certainly there were cavalry commanders of efficiency employed in the Peninsula –

Sir Stapleton Cotton, who commanded the force as a whole from 1810 to 1814, was solid and reliable, showing competence and skill on a number of occasions, while brigadiers such as Le Marchant at Salamanca (22 July 1812) and Baron Von Bock at Garcia Hernandez (23 July 1812) acted brilliantly – and others, such as Lord Paget, were available but hardly used. Yet Wellington seems to have preferred men who never questioned his orders or showed any inclination to act with initiative, a direct result of his desire to hold the cavalry in check at all times, and consequently he tended to employ nonentities. Some, like Sir William Erskine and John Slade, were clearly incompetent, while others, like Sir John Vandeleur and Sir Hussey Vivian, were so intimidated by the Duke's presence that they took nothing which might be construed as an unauthorised action. This particular problem, coupled with the rather haphazard organisation of cavalry brigades, which often included both heavy and light units together and rarely stayed the same for long periods, effectively destroyed much of the independence of action which cavalry required. Brigadiers were quite willing to wait for personal orders from Wellington himself in a battle, thereby losing many opportunities for decisive engagements if they had chosen to go it alone. In the pursuit after Salamanca, for example, George Anson, commanding the 11th, 12th and 16th Light Dragoons, lost a number of chances to close with the French because he was not sure of his precise instructions, while at Waterloo (18 June 1815) the two light cavalry brigades, under Vandeleur and Vivian, were not committed to the battle until very late in the day for similar reasons. In such circumstances the tactics to be employed by cavalry – and particularly light cavalry – were obviously restricted.

Taking all these difficulties together – the Peninsula terrain, lack of available regiments, transportation, remount and marching problems, and unenterprising commanders – it is extremely surprising to find that any light cavalry tactics at all were developed between 1808 and 1814. But developed they were, for although occasional units displayed deficiencies in adaptability, some of the better regiments settled down to Peninsula warfare, learnt to live with the problems, and improvised in the best traditions of the British army. As a result they acted usefully in a number of important roles, the exploitation of victory, the covering of retreat, even reconnaissance, skirmishing and outpost duty, although, inevitably, the emphasis remained firmly upon the charge.

Tactics 1808-1815 The Charge

In 1811, at the height of the Peninsula campaigns, William Müller, "Lieutenant of the King's German Engineers, ... and late First Public Teacher of Military Sciences at the University of Gottingen", published a three-volume work in London entitle *The Elements of the Science of War*. In the sections dealing with cavalry he specified three basic formations for use in the charge, and although his views were by no means official policy in the British army, it seems logical to suppose that they reflected prevailing military thinking. The first formation he described as "Charging in Line", which entailed the simultaneous advance of all portions of a cavalry unit against the enemy in extended line, accelerating from a walk, through a trot, to a gallop, "in order that the horses may arrive in wind and full vigour, when they shock with the opposing body". This formation, Müller opined, was equally applicable against infantry or cavalry, although he seemed to favour its use exclusively against the latter. Secondly, he specified "the Charge in Echelon", which may best be likened to the naval formation "line astern", for it entailed equal sections of the cavalry unit hitting the same area of the enemy line at precise intervals. This had the advantage of continuous shock action at one particular place, and was useful against infantry or cavalry in column or extended order. Finally, he discussed "the Charge in Echiquer", which involved a checkerboard formation of relatively small attacks against certain portions of the enemy, be they artillery, infantry squares or cavalry units, with support being sent to those areas only where success was achieved (See Figure 1 for diagrams.) In brief, the three formations may be described as extended shock action, concentrated shock action, and probing assault.

Such manoeuvres on their own, however, were not sufficient to ensure success, for that depended to an overwhelming extent upon certain important requirements which had to be met by the troops involved. Firstly, the cavalry units engaged in a charge had to be well led and well disciplined. They had to be restrained from breaking into a gallop too soon, Müller set the ideal distances at three hundred yards from opposing infantry and eighty to one hundred yards from cavalry. They had to maintain the tactical deployment chosen for the particular attack regardless of casualties and present a solid front to the enemy at all times. The attitude and example of the officers was of crucial importance here, for, as Müller pointed out, "the best troops will crowd and fall into disorder if badly led". Connected to this, as the second point, the attack had to be stopped as soon as success had been achieved or failure made obvious. If the troops were not rallied quickly for further attack, defence or withdrawal, they would split into small groups, chasing fugitives far and wide, and would easily succumb to the slightest opposition. Thirdly, whatever the formation employed, some form of reserve was essential, either to exploit existing success or to cover repulsed troops as they withdrew to safety. Fourthly, the type of formation to be used and the description of cavalry to be employed had to be carefully considered – after

Figure I Theoretical Cavalry Formations for the Charge

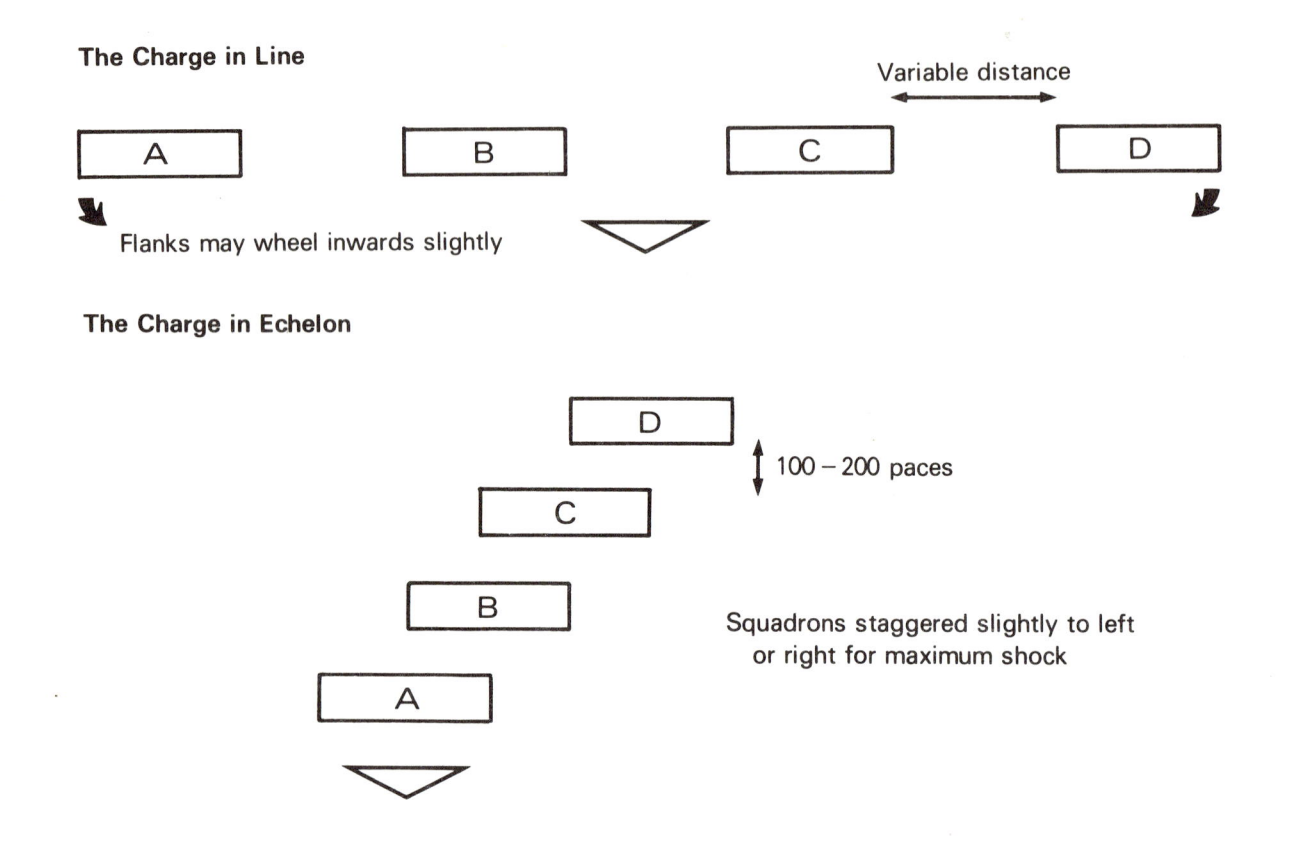

The Charge in Line

A B C D

Variable distance

Flanks may wheel inwards slightly

The Charge in Echelon

D
C
B
A

100 – 200 paces

Squadrons staggered slightly to left or right for maximum shock

all, charging in line against active artillery, or to a lesser extent, against steady infantry in square, would not achieve a great deal, while the employment of light against heavy cavalry would usually result in disaster. Finally, whenever a charge was ordered, the directions to the troops had to be intelligible and the ground to be covered reconnoitred for possible obstacles. It was no good pointing vaguely to the front shouting "charge" if the disposition, strength and type of enemy was unknown, nor was it advisable to advance at full gallop over ground which might contain watercourses, marshes or small rivers.

Taking such requirements together, it is obvious that the charge was not a tactic to be employed lightly, for it necessitated a significant degree of specialised training, military skill and basic preparation. Unfortunately there is little evidence to suggest that British cavalry units, either heavy or light, possessed these essential attributes during the

The Charge in Echiquer

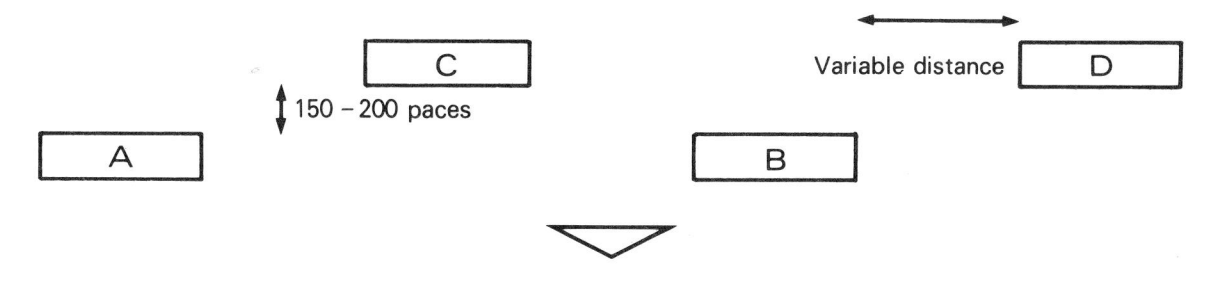

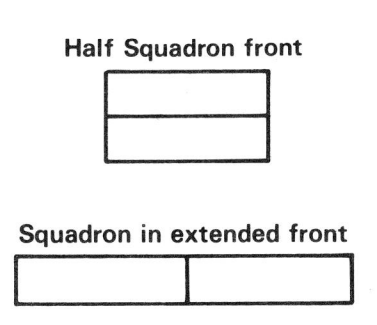

```
                    ┌─────┐                          ┌─────┐
                    │  C  │          Variable distance│  D  │
       ┌─────┐  ↕ 150-200 paces       ←────→         └─────┘
       │  A  │      
       └─────┘              ┌─────┐
                            │  B  │
                            └─────┘
                              ▽
```

Half Squadron front

Squadron in extended front

Peninsula and Waterloo campaigns. If Müller's formations were known, and there is nothing to prove even that degree of sophistication, they were apparently dismissed as theoretical and the problems of the charge consistently ignored. The main reason for this was undoubtedly a lack of specialised training, as Tomkinson of the 16th Light Dragoons noted: "In England I never saw nor heard of cavalry taught to charge, disperse and form, which if I taught a regiment one thing it should be that" – but the general attitude of the officers did nothing to help. This was summed up by Wellington in 1812 when expressing his annoyance over General Slade's inept handling of a heavy cavalry brigade at Maguilla. In his opinion such occurrences were "entirely occasioned by the trick our officers of cavalry have acquired of galloping at every thing, and their galloping back as fast as they gallop on the enemy. They never consider their situation, never think of manoeuvring before an enemy – so little that one would think they cannot manoeuvre, except on Wimbledon Common".

So far as the light cavalry were concerned, such ignorance and inefficiency, although inexcusable, is at least understandable. When the Peninsula campaigns began in 1808, the victories of the 15th Light Dragoons at Emsdorf and Villers-en-Cauchies hung like a shadow over the other light cavalry units. Emulation became the primary concern of most regimental officers, and, as both these victories seemed to have been achieved without the use of specialised formations, it became alarmingly acceptable to regard the full-gallop charge, in extended line, as the only manoeuvre worth effecting. In normal circumstances such a dangerous generalisation might well have been qualified through experience, but unfortunately the "lesson" was apparently reinforced, again by the 15th, in one of the first light cavalry engagements of the war.

In late 1808 the 15th, by now retitled Hussars, joined Moore's army in Spain, which was in the

Sahagun 21 December 1808

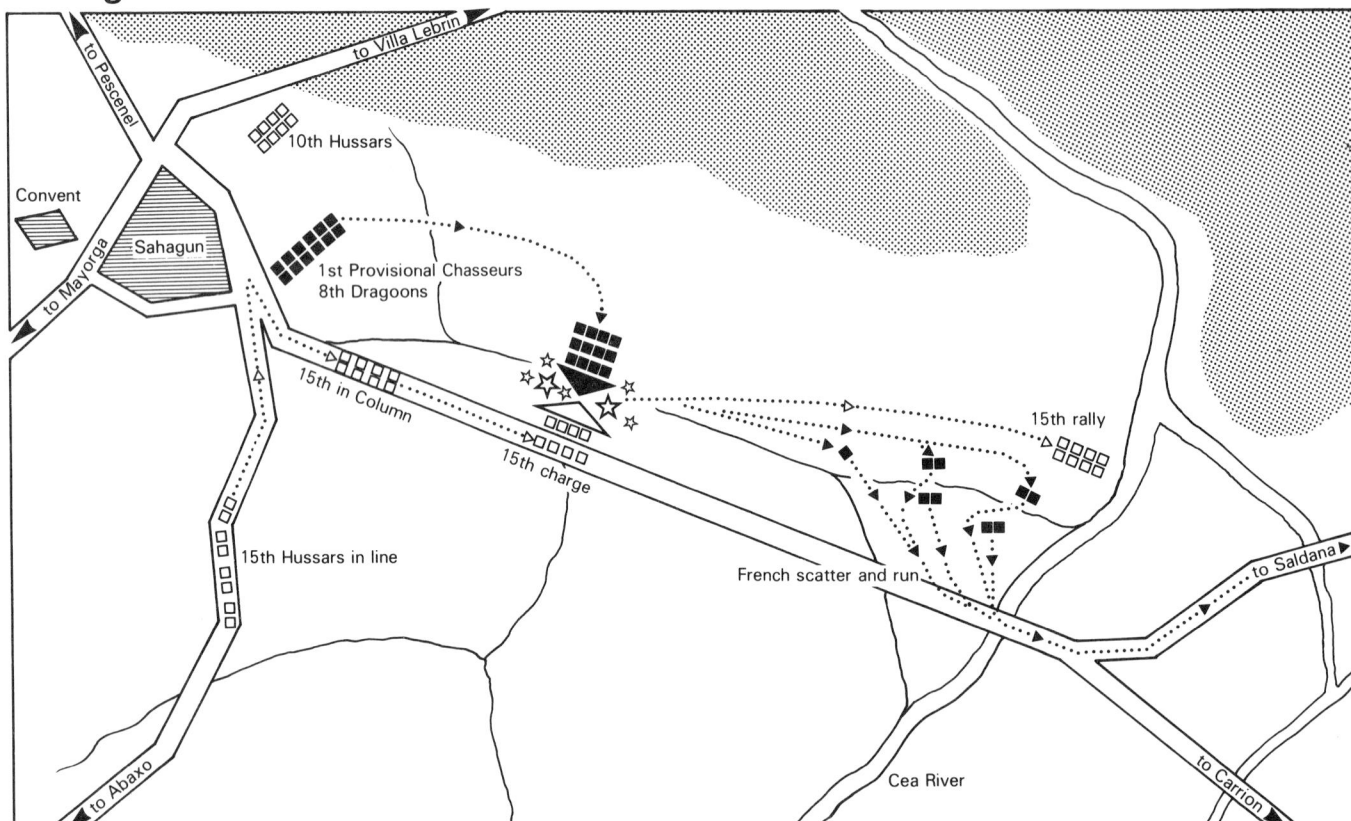

process of advancing north-eastwards from Salamanca towards Burgos. On 21 December the regiment, together with the 10th Hussars and a detachment of Horse Artillery, was sent in front of the main army to surprise a body of French cavalry and artillery in a convent at Sahagun, a large town on the Cea, a few miles from Melgar de Abaxo. The original idea was for the 10th and Horse Artillery to attack the convent while the 15th, under the British cavalry commander Lord Paget, circled round to cut off any enemy retreat. In the event, this did not work, for the French outposts were alerted, their two cavalry regiments, later identified as the 8th Dragoons and 1st Provisional Chasseurs, took to their horses, and a considerable force confronted the 15th alone. In the words of Captain Alexander Gordon of that unit, "the Fifteenth then halted, wheeled into line, huzzaed, and advanced". To shouts of "Emsdorf and Victory" the hussars charged full gallop, down-hill, into the French, who, after a short struggle, broke and fled. It was a decisive victory, costing the lives of twenty French cavalrymen and yielding over 170 prisoners, including two colonels and eleven other officers. Furthermore, it persuaded the French that Moore's cavalry force was stronger than was actually the

Talavera 28 July 1809

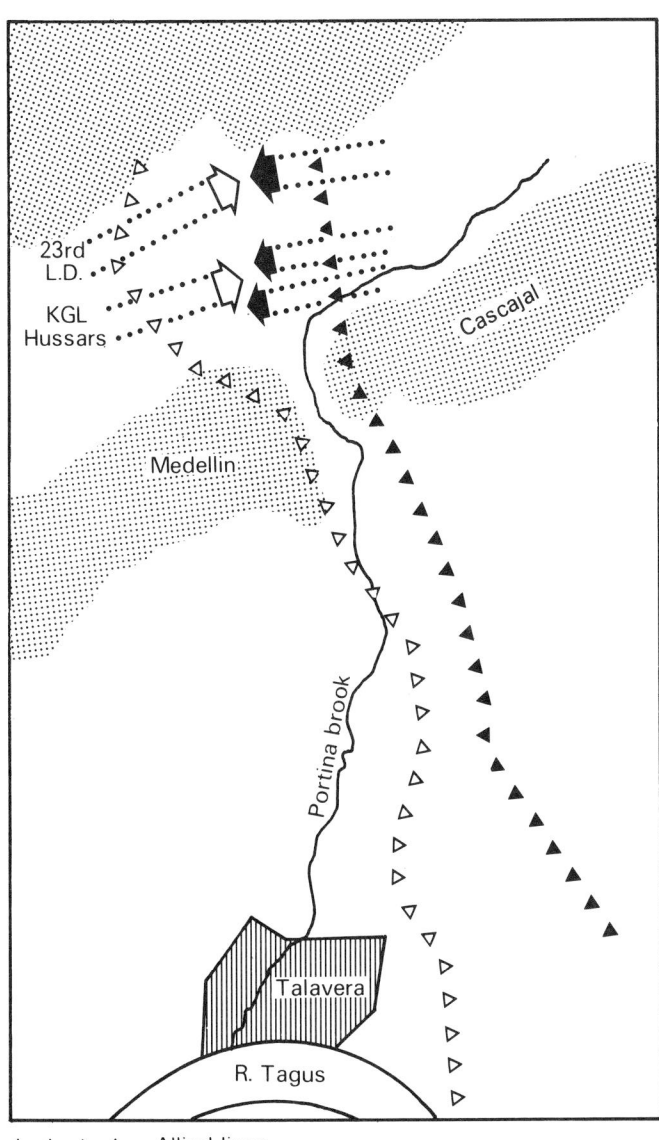

△ △ △ △ Allied lines
▲ ▲ ▲ ▲ French lines

case, making them extremely careful when attacking the British army on its subsequent retreat to Corunna. But from the point of view of tactical development in the light cavalry, the engagement was disastrous. Once again the 15th had enjoyed the success which other regiments craved, and, moreover, they had done so by means of the full gallop charge. No one chose to take note of the fact that the regiment had been expertly rallied by its officers, so preventing dissipation and disorder, and no one questioned the lack of a reserve in the charge, which could have been unfortunate. Instead, the action was seen as further proof of light cavalry capabilities in a tactic for which they had not been designed, and other units strove even harder to repeat the success. On the whole they proved incapable of doing so, and in fact achieved very little.

In many ways this is surprising, for light cavalry short comings in the charge were shown as early as Vimiero (21 August 1808), when two squadrons of the 20th Light Dragoons, after successfully cutting up a beaten column of French infantry, pushed on for half a mile in small groups, without rallying, to charge Junot's cavalry reserve. They were badly and deservedly maltreated, losing about a quarter of their effective men and horses, a loss which in fact ensured that the squadrons were of no further use in the campaign. But other units did not learn from this mistake: on 25 March 1811, for example, at Campo Mayor, the 13th Light Dragoons charged and defeated the 26th French Dragoons, capturing eighteen extremely valuable siege-guns being escorted to the fortress of Badajoz. The charge itself, delivered at full gallop, was well executed, but the regiment made no attempt to rally. Isolated detachments of light dragoons galloped on for more than six miles, sabring the scattered fugitives, stopping only when they came under direct fire from the Badajoz garrison. Meanwhile, the captured siege-guns had been left without a guard, enabling the

French drivers to take them back to safety. It was reported that Wellington was so incensed by this affair that he threatened to deprive the 13th of their remaining horses, sending the men home and using the animals as remounts for other regiments.

Such recklessness and indiscipline, which can only be blamed upon the officers, who gave no thought to the rally, could easily destroy a light cavalry regiment completely, particularly if it was combined with another deficiency. This was shown to good effect at Talavera (28 July 1809), where the extra deficiency was a lack of prior reconnaissance. A strong French attack had gone in against the British left and a fresh infantry division with two light cavalry units in attendance, was moving to support it. Wellington noted this and ordered General Anson's brigade, consisting of the 23rd Light Dragoons and 1st Hussars K.G.L., to charge the enemy reinforcements over what appeared to be a smooth and level plain. The initial deployment of these two light cavalry regiments was well-managed, but in the long advance towards the enemy the 23rd began to increase its pace. There was no reason for this, as the French were still several hundred yards away, but the officers made no attempt to check the tendency. As a result, the regiment was virtually at full gallop when it suddenly came upon an abrupt cleft in the ground hitherto unnoticed.

The leading officer just managed to spur his horse over this obstacle, which was in fact a dried streambed, eight feet deep and twelve to eighteen feet across, but the speed of advance gave him no time to warn the rest of the regiment. Horses and men fell in a confused and limb-breaking jumble in the middle of the ravine, the second line of horsemen, following too closely upon the first, crashed into them, and little more than half the regiment clambered up the other side. All formation disappeared, and although the remnants were rallied after a fashion and the advance continued, still at an uncontrolled pace, the damage had been done. Two squadrons passed between French infantry squares to rush headlong into a brigade of chasseurs half a mile away. By that time the 23rd was in great disorder, on spent horses, and outnumbered five to one. Despite support from the German Hussars, who had advanced at a more controlled speed and successfully negotiated the ravine, the regiment lost nearly half their strength – 105 taken prisoner and 102 killed or wounded – and were utterly useless for further campaigning. They had to be relieved by the 13th Light Dragoons from Britain, an exchange which, from the evidence of Campo Mayor, was not beneficial.

It could of course be argued that, despite the details of inefficiency, the Talavera charge achieved its aim. The enemy attack on the left was blunted and the French, no doubt bewildered by the sheer audacity of the 23rd, withdrew. But it was instances such as this which contributed to Wellington's distrust of the cavalry as an offensive weapon, and the three examples cited above – Vimiero, Talavera, and Campo Mayor – were in fact the only occasions upon which the Duke used the light cavalry charge in the Peninsula to affect the course of a major battle or campaign. He realised the deficiencies of the tactic, and chose to depend for success upon his infantry and artillery rather than mounted regiments which consistently displayed indiscipline, recklessness, poor leadership and bad tactical appreciation when called upon to attack the enemy in strength. This may seem a harsh judgement in retrospect, but at the time, it was more than justified. It goes far to explain the relative lack of light cavalry involvement in the Waterloo campaign – beyond the late advance of Vandeleur's and Vivian's brigades in the battle itself, the only incident of note was the charge of the 7th Hussars at Genappe on 17 June, which will be examined as a light cavalry tactic for covering a retreat – and made the evolution of other tactics in the Peninsula essential.

Tactics Pursuit and Retreat

Fortunately, however, other tactics were developed which, through a lack of alternative possibilities, necessarily referred back to some of the original functions of light dragoons. The first and potentially the most decisive of these was pursuit. Light cavalry units had been designed initially for speed, flexibility and manoeuvrability; as their professed predilection for shock action gradually proved impracticable, these characteristics made them ideally suited for harassment of a broken enemy and the exploitation of victories achieved by other arms. The fact that this was recognised between 1808 and 1815 is apparent from Wellington's statement, quoted already, that "when the infantry have beaten the French, then the cavalry, if they can act, make the whole complete".

Nevertheless, the operative phrase remains "if they can act", for as with so many aspects of cavalry warfare in the early nineteenth century, problems emerged which made the tactic of effective pursuit extremely difficult. Presuming a British victory to have been complete, with the French withdrawing in the kind of disorder which made pursuit advisable, the first problem was geographical. Under this heading, the most common consideration was the type of terrain to be covered. Many of Wellington's battles were fought in defensive positions overlooking broken ground which caused the enemy to split his forces, usually while attacking uphill. This had obvious advantages for the main infantry engagement was concerned, but once the possibility of pursuit arose, the cavalry inevitably experienced difficulties in swift or coordinated movement. At Busaco (27 September 1810), for example, Wellington fought from a ridge overlooking rough and irregular ground which precluded rapid cavalry deployment once the action was over. After Orthez (27 February 1814), although pursuit was ordered and some success gained, the horsemen found it difficult to move quickly over enclosed agricultural land. In addition, the weather could have an adverse effect upon pursuit. If it rained, as at Albuera (16 May 1811) and throughout much of the Waterloo campaign, the cavalry suffered considerably from a loss of secure footing in the mud. Finally, if an engagement continued into the evening it was advisable not to let the cavalry loose, for once darkness fell chaos was sure to result, a consideration which contributed to the lack of immediate exploitation after Salamanca.

But even if the ground, weather and light were favourable, there were often purely military reasons why pursuit could not be ordered. In some engagements, for instance, Wellington just did not have sufficient cavalry forces at his disposal with which to harry the French. At Rolica (17 August 1808) he had a mere two squadrons of the 20th Light Dragoons, desperately short of horses, whereas after Vimiero his light cavalry had been incapacitated through a poorly executed charge during the battle itself. On other occasions Wellington was loath to initiate movement of any kind from the battle area because of the weak state of his infantry. At both Talavera and Salamanca the Spanish troops under the Duke's command displayed alarming deficiencies of morale and fighting ability which, combined with British infantry losses, obliged him to keep his army together rather than risk a dissipation of strength through pursuit. As a final military factor, if there was chance of plunder, the dispatch of any forces under independent command was asking for trouble, as they were more likely to chase loot than the main enemy army. After Vittoria (21 June 1813), for example, when pursuit was ordered, elements of the 18th Hussars were caught in the act of plundering when they should have been exploiting the victory. According to Tomkinson, "Lord Wellington was so much enraged that he would not recommend any of their subalterns for two troops, which were vacant by two captains killed, a thing very unusual."

All in all, therefore, the tactic of pursuit was difficult to organise and could be used on very few occasions. In fact, taking the Peninsula and Waterloo campaigns as a whole, there were only four instances upon which light cavalry managed to harass a defeated French army. Even then, the problems were by no means over. In one of the actions a certain degree of localised success was achieved – after Orthez, despite the terrain, the 7th Hussars managed to cut off and take prisoner about 2,000 stragglers in a running fight which lasted until nightfall – but in the others it was apparent that the troops involved were remarkably incapable of exploiting their opportunities. As with the charge, once the regiments had surmounted the multifarious physical or general difficulties and manoeuvred into a situation where they could act, deficiencies of leadership, discipline and military skill prevented complete success. This was seen as early as May 1809 when an advance squadron of the 14th Light Dragoons found itself in an ideal position to harass the French forces demoralised by Wellington's sudden crossing of the Duero at Oporto. Instead of cutting off stragglers and pushing the enemy to the point of panic, the light horsemen involved themselves in a reckless charge over difficult ground against the most solid part of the French rear-guard. They managed to defeat the enemy, but at such a cost, thirty-five men killed or wounded out of a total of 110, that further pursuit was impossible. A similar action took place after Vittoria when two squadrons of the 12th and 16th Light Dragoons attacked a superior force of French cavalry in line, for although the casualty rate was not so high, the British troopers were repulsed and the pursuit quickly petered out.

These charges, however, were at least attempts at exploitation, with opportunity targets being attacked in the only way known to light cavalry units of the time. Almost the opposite happened after Salamanca, when Anson's brigade (11th, 12th and 16th Light Dragoons) caught up with the rear of Marmont's retreating forces. As soon as the enemy was sighted, far from a reckless charge being ordered, the brigade was halted and a mere three of the twelve squadrons available sent forward to assess the situation. This gave the French ample time to organise a screening force of skirmishers, which held the ground while the main body withdrew. By the time Anson realised his mistake and brought up the rest of the brigade, together with a detachment of horse artillery, it was too late and the French had escaped. As a participant in the action stated "General Anson here missed a good opportunity of doing something with (the French) rear . . . . It does not look like a quick advance following up a great victory, and I think they will be let off too easily." In the event, this proved to be the case.

Such deficiencies in leadership – shown both by the recklessness of the 14th Light Dragoons and the excessive caution of Anson – justify a conclusion that, despite their potential in the tactic, British light cavalry units did not operate well in pursuit. The problems they faced were immense and largely outside their control, it is true, but when the opportunities arose, as they undoubtedly did in the four above examples, no decisive results ensued. Much of the blame for this must lie in the lack of specialised training and military skill which, as was seen with the charge, characterised light cavalry units in the Peninsula, although once again it was probably Wellington's distrust of his horsemen as a whole which prevented more general success. He realised the defects of leadership and discipline and, as a contemporary pointed out, did "not like to entrust officers with detachments to act according to circumstances" for fear of the inevitable consequences. It is hardly surprising to note, therefore, that the light cavalry were regarded, even by some of their own officers, as "rather deficient in the pursuit of a broken enemy."

But the adaptability of light cavalry, which made their involvement in exploitation theoretically advisable, could be put to other uses. During the campaigns under study the British army was not always winning great victories; on some occasions it was out manoeuvred or out fought and forced to retreat. Whenever this happened, the infantry and foot artillery, to say nothing of the commissariat and support services, were extremely vulnerable. Their morale was low, they understood little about their situation, and they gradually lost order as they moved slowly away from the French. In such circumstances they naturally needed protection, particularly if the enemy was pushing hard, and it was here that the light cavalry could be especially useful. As Müller pointed out in 1811, "cavalry is principally required to cover the retreat of infantry in open . . . . ground", and for once this appears to have been fully recognised in the mounted regiments. Success was achieved on a number of crucial occasions.

The first and by far the most impressive of these was during Moore's retreat to Corunna in the winter of 1808-9. After the victory of the 15th Hussars at Sahagun, preparations were made for a general advance on Burgos. However on 23 December 1808 Moore received intelligence that fresh enemy forces were moving towards him from Madrid and that Napoleon himself was trying to cut his lines of communication by advancing on Benevente. Thus in danger of being out-manoeuvred, the British advance was countermanded and a full scale retreat on the port of Corunna immediately ordered. Almost at once the cavalry under Paget, consisting entirely of light units, moved in the direction of Carrion and Saldana while the infantry fell back on Valencia and Valderas. When the retreat began on 24 December, in atrocious weather conditions, the cavalry took the rear-guard, hoping to gain time for the rest of the army by delaying the French advance. They started off well, for the initial demonstrations coupled no doubt with the memory of Sahagun, caused the enemy to over-estimate the strength of Paget's force and to move with more caution than was in fact necessary. This gave the infantry a head-start of over twenty-four hours, enabling them to reach Valderas unmolested, but as the British plan became clear, the French advanced with speed. Elements of Soult's cavalry caught up with the rear-guard late on Christmas Day, and from then until 31 December, when Moore reached the comparative safety of the Galician mountains beyond Astorga, the British light cavalry units bore the brunt of the French attacks. Their record of success was impressive, for it is more than justifiable to claim that without their continuous protection the bulk of the army would not have been saved.

French attempts to destroy Moore's army took two forms. While Soult's advance units pushed the rear-guard, trying to break through the protecting screen, detachments of his cavalry moved across country, outflanking Paget's force, and cut the road of retreat. This meant that on a number of

Light Cavalry 1796-1806

Corporal
10th Regt.

Trumpet-Major
1796
10th Regt.

Trooper
c.1800
17th Regt.

Trooper in full dress 1796
16th Regt.

Trumpeter in full dress c.1800
17th Regt.

Trooper in full dress 1790
16th Regt.

Officer in undress 1790
10th Regt.

Colonel
1795-1805
16th Regt.

Officer
1805
16th Regt.

Colonel
1802
10th Regt.

Officer in service dress 1806
20th Regt.

Officer in full dress 1805
14th Regt.

Officer in service dress 1800-1805
12th Regt.

Officer in full dress 1805
7th Regt.

occasions the British rear elements had to face two ways, and the only tactic which could be used in such circumstances was the charge. Fortunately, under Paget's expert leadership, this was, for once, well-executed, displaying none of the usual characteristics of recklessness or indiscipline. As early as 26 December, for example, the 10th Hussars, with the ubiquitous 15th in support, successfully charged two squadrons of French chasseurs who had cut the road at Mayorga. Less than twenty-four hours later, on the retreat to Benevente, the 18th Hussars turned on their pursuers a total of six times, on each occasion charging so well that they were left unmolested for the next few miles. Incidents such as these added to French confusion about the strength of the British cavalry, as late as 1 January 1809 Napoleon estimated their numbers at "4,000 or 5,000 horses" when in fact they mustered barely half that total, and forced them to slow their advance. This in turn took pressure off the retreating army, enabling the infantry, by now in considerable disorder, to cover a few more miles to safety.

When they were not being directly attacked, however, the cavalry was by no means idle. As the army retreated, attempts were made by the engineers to destroy important river bridges, and while the explosives were laid, Paget's horsemen endeavoured to keep the enemy at bay. At the same time detachments were sent to destroy any ferries which might exist and picquets were placed at likely fording places to prevent outflanking moves. Once the bridge had been blown, the retreat continued, although it was common policy to leave the ford-picquets in place for a while to warn of any sudden advance. Sometimes, if the French came on too soon and the bulk of the cavalry was still in the vicinity, these picquets could even act as the bait in a trap, drawing the enemy into hurriedly but carefully laid ambushes. This happened dramatically at Benevente on 29 December. The bridge over the Esla, a fast-flowing river swollen by winter rains, had been destroyed, ferries in the area had been sunk, and the only ford was covered by a picquet of the 18th Hussars. French chasseurs of the Imperial Guard, commanded by General Lefebvre-Desnouettes, saw an opportunity for quick advance, crossed the river and attacked the picquet. Immediately the rest of the 18th, with the 3rd Hussars K.G.L. in attendance, charged into the assault, to be joined by elements of the 7th and 10th Hussars under Paget himself. The chasseurs were soundly beaten and pushed back across the river, reputedly under the furious eyes of Napoleon, and Lefebvre-Desnouettes was captured. Once again the retreat continued unmolested for a few more precious miles.

These tactics of charge and ambush were necessarily evolved in the field under the pressure of enemy attack, for no specific training in them had ever been carried out before 1808. The cost in men and horses was heavy. By the time the mountains were reached Paget's regiments were so weak that, with the exception of the 15th Hussars, who continued to cover the retreat through the defiles of Galicia, they were all sent immediately to Corunna for embarkation. But now the light cavalry had discovered a particular duty at which they could excel. This was not merely fortuitous, for on two occasions during Wellington's Peninsula campaigns – the withdrawal to the lines of the Torres Vedras in late 1810 and the retreat from Burgos exactly two years later – the tactics were repeated, and again proved successful. Unfortunately the record was rather tarnished in 1815, dring the Waterloo campaign, even though Paget, by now the Earl of Uxbridge, once more commanded the cavalry.

After the affair at Quatre Bras, in which the cavalry took no part, and the defeat of the Prussians at Ligny, both on 16 June, Wellington decided to pull his army back to a previously-chosen spot on the line of Mont St Jean. As soon as the infantry

The Corrunna Retreat

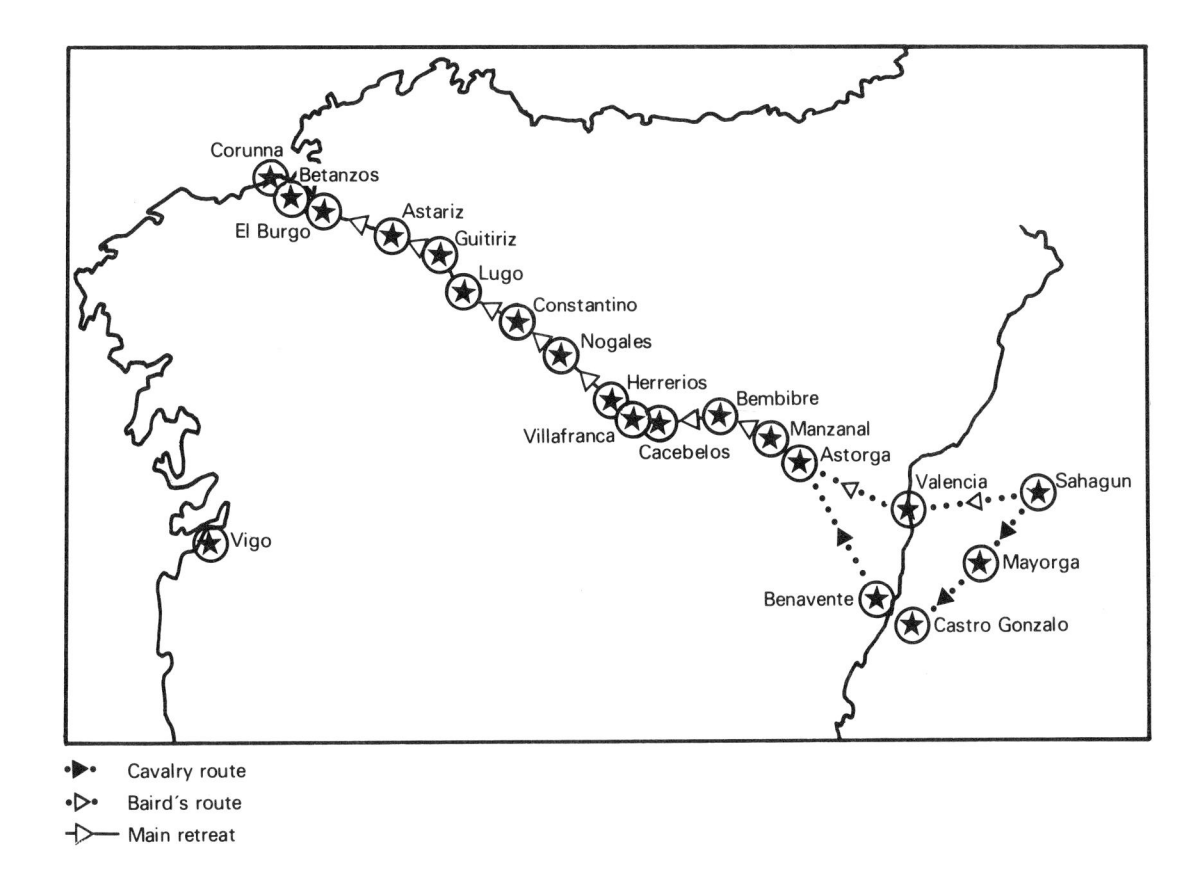

- •▶• Cavalry route
- •▷• Baird's route
- ─▷─ Main retreat

began their withdrawal on the morning of 17 June, the cavalry moved to the rear, with the 7th Hussars, 11th and 23rd Light Dragoons protecting the whole. At first all went well. Ney did not receive orders from Napoleon to advance until midday, two hours after the British retreat had begun, and when his cavalry caught up with the rear-guard they were kept at a distance by the 7th Hussars, who withdrew slowly by squadrons and repeatedly threatened to turn and charge. Even so, when the two opposing forces reached Genappe, a village two miles north of Quatre Bras, the forward elements of Ney's Polish Lancers were close behind the British.

Genappe was tactically of the utmost importance on the line of retreat, for through it ran the only road to Waterloo. North and south of the village this road ran along a raised causeway several feet above the fields, and the retreating army, forced to

29

Trumpeter in service dress 1796-1806
15th Regt.

Troop-Sergeant-Major
1796-1806
14th Regt.

Trumpeter
1796-1806
18th Regt.

Corporal full dress
1796-1806
7th Regt.

Sergeant in full dress 1796-1806
13th Regt.

Trooper in service dress 1796-1806
14th Regt.

Farrier in full dress 1796-1806
15th Regt.

30

Officer
1806
15th Regt.

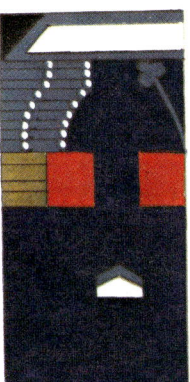

Officer
1806
11th Regt.

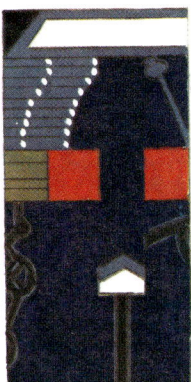

Officer
1806
9th Regt.

Officer in service dress 1806
9th Regt.

Officer in full dress 1805
18th Regt.

Officer in service dress 1804
10th Regt.

Officer in full dress 1806
13th Regt.

31

Waterloo 1815

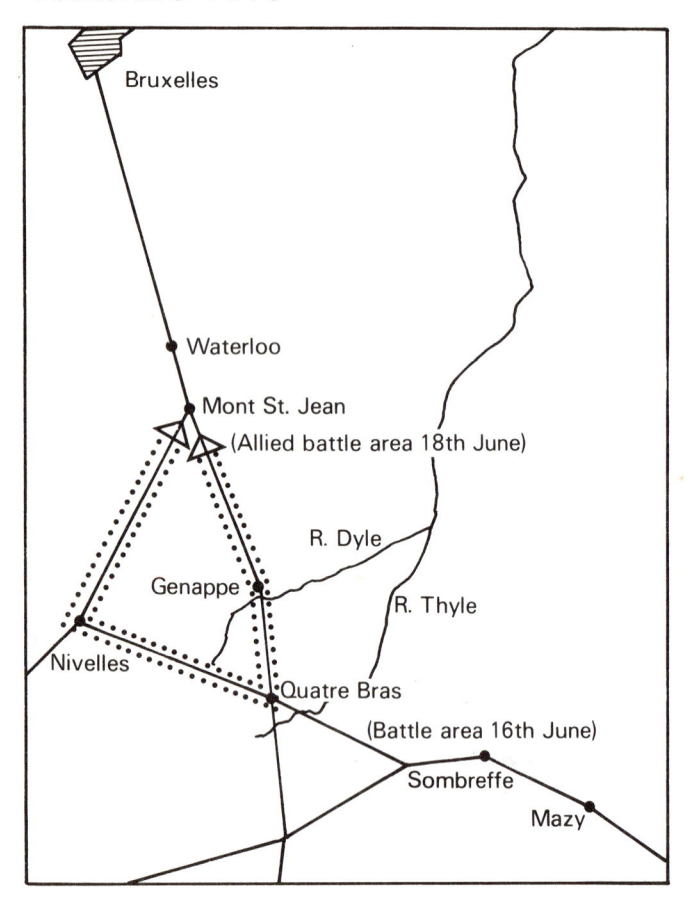

keep on this, presented a tempting target for cavalry attack on the flanks. Uxbridge therefore had no choice but to hold Ney's advance in the village itself, where the single narrow street forced the enemy into a bottle-neck which was extremely vulnerable to counter-attack. Indeed, as the lancers – "very young men, mounted on very small horses" – moved forward, they were unavoidably crammed into a column with a frontage of six men only, and so closely followed by other units that they could not retire if attacked. Until they deployed out of the village into the good galloping-country beyond, they were in a trap, presenting an ideal target. But this was not a target for light cavalry, who lacked both the weight and the arms to be effective, and for this reason it is surprising to find that Uxbridge chose the 7th Hussars for the assault. The fact that he was their regimental Colonel undoubtedly affected the decision, and the men charged willingly. But to little avail. As soon as they saw the hussars advancing, the lancers halted, dressed their ranks and lowered their lance-points. When the impact occurred it was, in the words of one of the British officers involved, like charging a house. Hussar troopers were pinned like beetles to a card, they lost nearly all their officers, and had little appreciable effect upon the situation. They were eventually driven back with heavy losses, the leading lancers close on their heels. Uxbridge then contemplated compounding his error by committing the 23rd Light Dragoons, but they, rather understandably, displayed little enthusiasm. It was not until the 1st Life Guards – "big men on big horses" – advanced to the charge that the French retreated back through and out of Genappe. The enemy learned their lesson, making no more efforts to interfere with the British withdrawal, but the apparent invincibility of light cavalry in retreat protection had suffered a serious setback.

This particular action was to have far-reaching consequences, for when the Napoleonic Wars ended and light cavalry returned to peace time soldiering, the idea of retreat-protection as a definite tactic was discounted on the evidence of Genappe. This was unfortunate, for the outstanding successes of 1808, 1810 and 1812 should have shown how well suited the regiments were for this role, and how useful it was to have fast moving, adaptable mounted troops capable of something more than ineffective and inefficient shock action.

Tactics Picquets, Reconnaissance and Skirmishing

A similar picture emerges with the last group of discernible light cavalry tactics. The light-dragoon advocates of the eighteenth century regarded reconnaissance, skirmishing and outpost duty as the most important roles of the new units. They stressed the need for observation and the gathering of intelligence in the field, the necessity of "distant advanced posts" to give early warning of enemy operations. Small fighting patrols were also effective to harass the enemy, and prevent him from gathering forage or moving freely in any spaces which might exist between or on the flanks of the opposing armies. Müller followed this up in the early-nineteenth century, devoting considerable portions of his work on cavalry tactics to detailed instructions regarding advanced posts, skirmishing and reconnoitring, and echoing his predecessors in the opinion that "light cavalry are well suited to undertake and accomplish" all such tasks.

Superficially, these theoreticians were more than justified in their views, even as late as 1811. When light dragoons were first raised they were designed specifically for mobility, flexibility and manoeuvrability under almost any conditions, and although the uses to which such troops were put in the late eighteenth and early nineteenth centuries went beyond those originally envisaged, these basic characteristics remained. They were reflected in the equipment which continued to be issued to the growing number of light cavalry regiments during the French Wars. Horses were smaller and tougher than their counterparts in the heavy units, being chosen especially for their ability to work hard and move quickly over difficult terrain. Obviously it would have been foolish to slow these animals down unnecessarily, so the men were generally of smaller stature and lower weight than dragoon guard or dragoon troopers. Indeed, as early as the 1780's it had become accepted policy for light cavalrymen to be transferred to heavier regiments once they surpassed a certain height or weight (usually five feet eight inches or twelve stone). For the same reason saddles, bits and other items of horse furniture were lightweight, following closely ideas first put forward by the Earl of Pembroke in 1761.

It was the uniforms and weapons of the individual troops, however, which reflected most clearly the continuing possibility of a return to original functions. When the Peninsula campaigns began, those regiments which retained their light dragoon status wore uniforms which had altered little since the 1780's, when the lessons of the American War of Independence were still fresh. Black japanned helmets with a bearskin crest along the crown – known as the "Tarleton" helmets after their originator – blue, loose-fitting coats with white froggins, and buckskin breeches over Hessian boots, combined to provide comfort and durability, ideally suited to the rigours of sustained patrolling and bivouacking in the field. Their only disadvantage was the lack of protection from sword cuts, but, since they had been designed at a time when the outpost duty was seen as more important than the charge, this is understandable.

Unfortunately, these sensible uniforms were on

Officer   Officer   General-Officer of Hussars
1808-1812   1808-1812   full dress
7th Regt.   10th Regt.   1807-1812

                                                         Officer in marching order 1808-1812
                                                                      7th Regt.

Officer in review order 1805-1807   Officer in marching order 1808-1812   Officer in service dress 1808-1812
10th Regt.   18th Regt.   15th Regt.

34

Hussars 1807-1812

Troop-Sergeant-Major
1810
18th Regt.

Trumpet-Major
1810-1812
7th Regt.

Sergeant full dress
1811-1813
10th Regt.

Trooper in service dress 1808-1812
18th Regt.

Trooper in service dress 1807-1809
10th Regt.

Trooper in marching order 1808-1812
7th Regt.

Trooper in full dress 1808-1810
15th Regt.

35

the way out in 1808. The hussar regiments, had already discarded comfort and serviceability for extravagance and show. Their fur busbies, with brightly-coloured busby bags, were top heavy, insecure and, as was discovered during the Corunna campaign, alarmingly liable to disintegrate when wet, while their unwieldy pelisses and tight pantaloons made arduous service an uncomfortable prospect. Slight modifications were made in August 1812, when the busbies were replaced by the more protective and hard-wearing cylindrical pill-box hats, but the advantages were effectively cancelled out in the cavalry by dramatic changes to light dragoon uniforms. These substituted a bell-topped shako for the Tarleton helmet, the blue jackets were replaced by shorter and tighter blue coats, and the buckskin breeches gave way to close-fitting trousers made of webbing. The protective qualities of these new items were slightly better than their predecessors, but not enough to warrant any improvement in the change, while the serviceability of the old uniforms had been lost completely. Nevertheless, for much of the Peninsula period the light dragoon regiments under Wellington's command were clothed in a way ideally suited for reconnaissance, skirmishing and outpost work.

By 1808 the weapons carried by private troopers throughout the light cavalry suggested that something more than shock action might be expected. The 1796 pattern sabre, for example, although undoubtedly of use in the mêlée of the charge, was designed for cutting rather than thrusting, a movement applicable to the more independent actions of the skirmish. Furthermore, each trooper had a flint-lock pistol in a saddle-holster and a smooth-bore carbine, "manufactured in accordance with the general principles of the Baker rifle", attached to his waist-belt by means of a metal ring. Both weapons were obviously long-range when compared to the sword, requiring time for reloading and stability for aiming which the speed and confusion of the charge did not provide.

Taking all these advantages of equipment and weapons together, it might well be imagined that the duties of guarding the main army and observing the enemy's movements in the field would have been fully recognised in the light cavalry regiments. But they were not. As has been intimated already, once the 15th Light Dragoons had shown the way at Emsdorf and Villers-en-Cauchies, nothing except the charge was regarded as important. Light cavalry officers craved the glory of shock action and naturally regarded such tasks as reconnaissance or outpost duty as boring and mundane, if not a little beneath their dignity. This was shown in their attitudes to pamphlets and books which outlined the theoretical detail of such duties, for although this was a period of scanty tactical thought, a number of relevant works did exist. Simes and Hinde, for example, were still available, while Dundas' *Rules and Regulations*, which every cavalry officer had to procure by Royal Command, contained a section on skirmishing and the gathering of intelligence. Beyond these, Le Marchant wrote a pamphlet on reconnaissance in 1798, and in the same year Major-General Baron de Rottenberg's *Regulations for the Exercise of Riflemen and Light Infantry,* dealing in detail with outposts, reconnoitring and skirmishing, was specifically recommended to the light cavalry by the Adjutant-General. There was therefore no excuse for ignorance, but, inevitably, it existed. As late as March 1812 Tomkinson could write that "to attempt giving men or officers any idea in England of outpost duty was considered absurd", adding that on the only occasion of which he was aware when training had been carried out before 1808 the cavalry commander had "got the (supposed) enemy's vedettes and his own looking the same way".

In circumstances such as these, where the obvious

potential of light cavalry was ignored and particular duties treated with disdain, any attempts at reconnaissance, skirmishing or advance posts were sure to meet with small success. When the 15th Hussars joined Moore's army in 1808, for example, their Adjutant was free to admit that "the outposts were in general harassed by their want of knowledge in taking up proper positions". Important geographical features were not guarded, vedettes were easily picked off by the French, mutual support between posts or patrols was almost nonexistent, and intelligence gathering was poor. Nor was this merely a passing phase in the general developments of skills, for as late as 6 April 1811 an entire squadron of the 13th Light Dragoons was surprised and captured by the enemy while on the outpost duty near Elvas. Two months later a squadron of the 11th Light Dragoons suffered a similar fate in the same area. According to Wellington, the latter incident tended "to show the difference between old and new troops", for the 11th had only just arrived from England. The truth was however that detached duties had not been trained for and so were sure to create casualties as the lessons were forcibly learnt under combat conditions.

Fortunately for the reputation of the light cavalry, individual units displayed an ability to improvise, and adapted to these important duties with commendable speed and skill. In fact the way was shown by the 1st Hussars, K.G.L., who laboured under none of the disadvantages caused by the British predilection for shock action, and were consequently better prepared to carry out almost any task presented to them. This unit, under their brilliant commander Frederick von Arentschildt, settled down to detached duty as soon as they arrived in the Peninsula in 1809, and their success rapidly became known throughout the army. Captain John Kincaid of the 95th Rifles summed up their reputation when he said that "if we saw a British dragoon at any time approaching in full speed, it excited no great curiosity among us, but whenever we saw one of the first hussars coming on at a gallop it was high time to gird on our swords and bundle up". This judgement was no doubt fully justified, but it would be unfair to presume that no British units contributed to the tactics of reconnaissance and protection. One has only to read the memoirs of such light cavalry officers as Alexander Gordon of the 15th Hussars or William Tomkinson of the 16th Light Dragoons to see that, in the regiments at least, there were continuous improvements between 1808 and 1815.

Reconnaissance was clearly essential in any military operation, for the army commander needed to discover all he could about the enemy's movements or intentions. Wellington depended to some extent upon spies for this service when planning his Iberian campaigns, but on the whole it was the light cavalry who provided the necessary intelligence. Strong patrols, usually consisting of about thirty mounted men under one or two commissioned officers, were sent towards the enemy lines, moving from village to village observing, taking prisoners and sending information back to the main army. Sometimes the patrol would be split, with portions watching specific roads or geographical features. On 22 October 1810, for example, Tomkinson noted that "we have one officer's picquet from the brigade on the Cadaval Road, one sergeant looking to the road up the hills, and one more to our left on the Obidos road. Our patrols", he added, "go up the hills, and for a league on their tops, to ascertain that (the enemy's) camps are in the place they have for some time occupied, and that no considerable body passes our way". This would appear to have been a fairly typical arrangement, and one that was repeated with regularity throughout the Peninsula campaigns.

Such mundane and repetitive work was not

Light Dragoons 1807-1812

Trooper in campaign dress 1808
14th Regt.

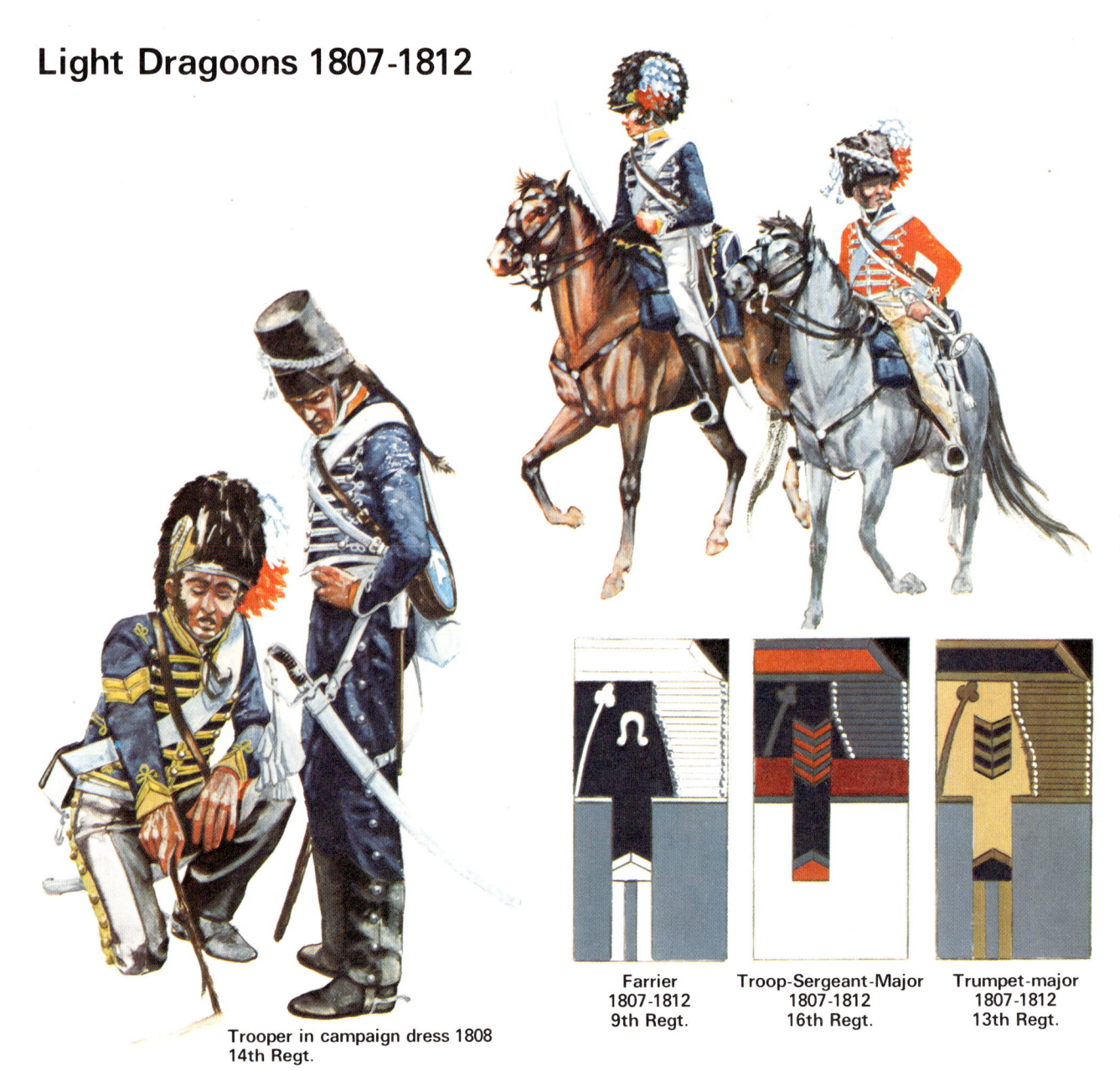

Corporal in service dress 1808-1812
13th Regt.

Farrier
1807-1812
9th Regt.

Troop-Sergeant-Major
1807-1812
16th Regt.

Trumpet-major
1807-1812
13th Regt.

Trooper in full dress 1807-1812
12th Regt.

Trumpeter in campaign dress 1808
18th Regt.

38

Officer
1808
14th Regt.

Officer
1808
9th Regt.

Officer full dress
1808
21st Regt.

Officer in service dress 1808-1812
23rd Regt.

Officer in marching order 1807-1812.

Officer in campaign dress 1807-1812
13th Regt.

Officer in service dress 1807-1808
16th Regt.

39

Figure II Light Cavalry Outposts 1808-1815

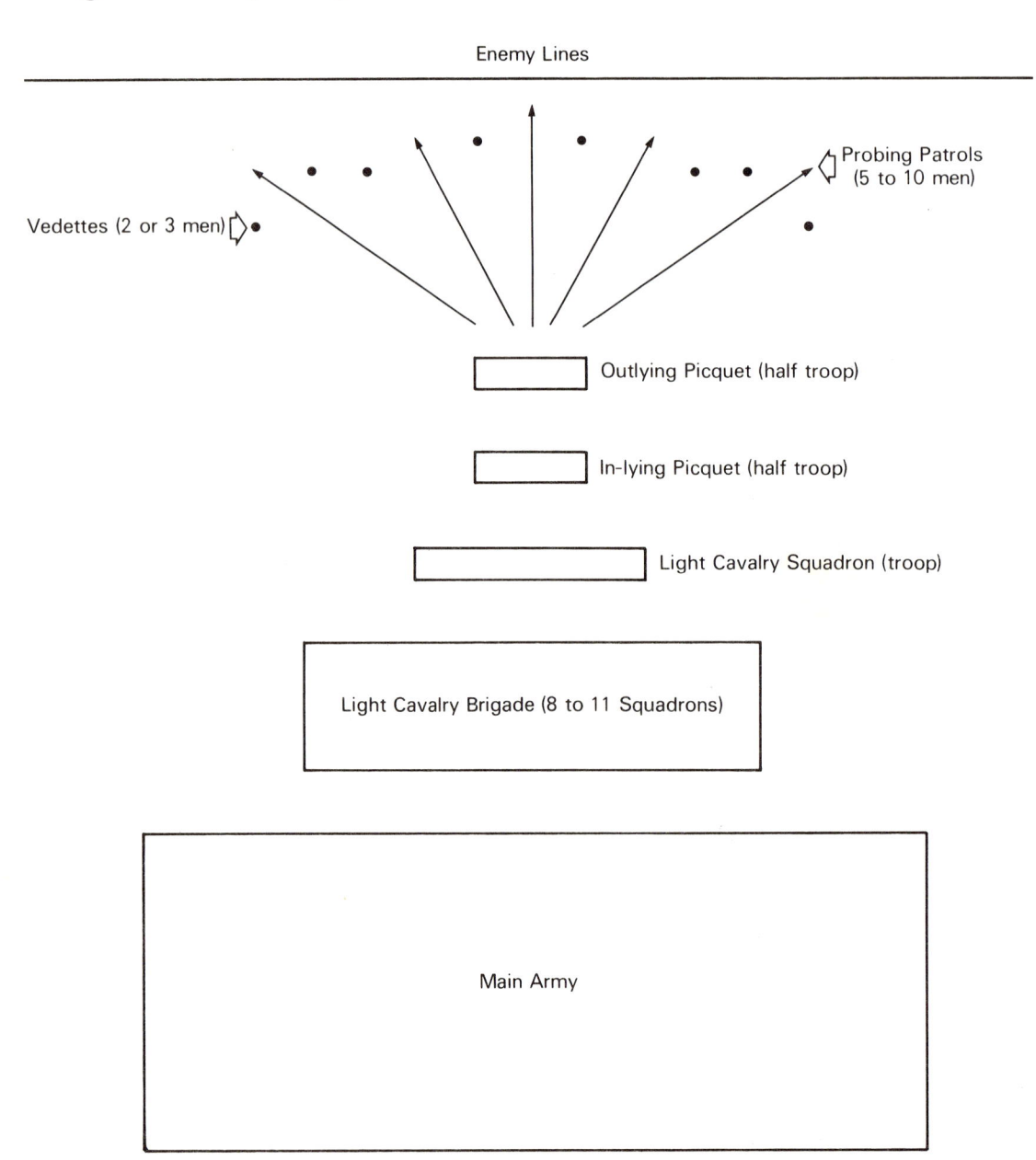

Enemy Lines

Vedettes (2 or 3 men)

Probing Patrols (5 to 10 men)

Outlying Picquet (half troop)

In-lying Picquet (half troop)

Light Cavalry Squadron (troop)

Light Cavalry Brigade (8 to 11 Squadrons)

Main Army

without its drawbacks for although there is no evidence of light cavalry patrols committing the cardinal error of attacking the enemy forces they were supposed to be observing, the accuracy and amount of information collected was occasionally very inadequate. This came to a head on 7 May 1811, when Wellington was forced to issue detailed General Orders on the subject, directing that:

> when an Officer makes a report of the movements of the enemy, he will specify whether consisting of cavalry, infantry, or artillery; the number, as far as he could judge; the time when seen, and the road on which moving; from what place, and towards what place, if the Officer can state it; and if reference should be made to the right or the left, in the report, care should be taken to state whether to the right (or left) of our own army or that of the enemy.

Generally however in the campaigns of 1808 to 1815, this appears to have been a relatively minor difficulty, more than cancelled out by the sustained record of light cavalry success in the tactic of reconnaissance. As one of the officers involved commented, patrols of this nature were generally regarded to be "of great service by way of gaining information", and since Wellington came to depend more and more upon his light cavalry it seems that they had found another duty at which they could excel. Indeed, some officers became so adept at reconnaissance as the campaigns progressed that the Duke sought them out especially for particular tasks. Major Charles Somers Cocks of the 16th Light Dragoons, for example, was used "confidentially, and constantly (up to his death at Burgos) in gaining intelligence of the enemy; for which purpose he frequently remained out days together at the head of thirty dragoons, working on (the French) flanks and rear".

There were similar successes when the light cavalry went on outpost duty and skirmishing. Whenever the army was stationary or contemplating movement it was desirable that it should be neither surprised by the enemy nor open to their observation. For this reason selected troops were stationed in front of the main body as a protective screen. Sometimes heavy cavalry or line infantry units performed this duty in the Peninsula, but normally it was given to light troops, both cavalry and infantry, since their composition provided the flexibility and adaptability which this service required. They would be stationed in places where the enemy could be observed, and the idea was for them to prevent surprise attack by acting as a buffer, holding or blunting the assault while giving warning to the rest of the army. At first, as has been mentioned already, this duty was not carried out with particular success, but after a remarkably short time a definite pattern of tactical behaviour emerged. When the army took up a position, a light cavalry brigade, usually with riflemen or light infantry in attendance, would be pushed forward, detaching a squadron to act as the outpost force. This squadron would then detach a troop which, when divided into two parts, would form what were known as the in-lying and out-lying picquets, the latter of which was nearer the enemy. From the out-lying picquet vedettes, consisting of two or three men each, would advance to observe the enemy or any likely avenues of approach, while probing patrols, consisting of anything from five to ten men, would move out in front and on the flanks to prevent surprise. (See Figure 2 for diagram).

In the event of an attack, the plan was for the patrols to give the first warning by galloping back towards the vedettes. They would then warn the out-lying picquet according to a series of previously arranged signals:

> When the enemy appeared, the vedette put his cap on his carbine. When he only saw cavalry, he turned his horse round in a circle to the left;

Hussars 1813-1815

Corporal in service dress 1811-1814
10th Regt.

Trooper in service dress 1807-1815
18th Regt.

Trooper in service dress 1813-1815
15th Regt.

Corporal
1815
7th Regt.

Trooper
1812-1813
7th Regt.

Sergeant
1812-1815
7th Regt.

Trooper in service dress 1814-1815
10th Regt.

Officer
1807-1815
18th Regt.

Colonel full dress
1814-1815
7th Regt.

Officer
1812-1813
7th Regt.

Officer in marching order 1813-1815
15th Regt.

Officer in service dress June 1815
10th Regt.

Officer in full dress June 1815
15th Regt.

Officer in full service dress 1814-1815
7th Regt.

43

when infantry, to the right. If the enemy advanced quick, he cantered his horse in a circle, and if not noticed, fired his carbine. He held his post until the enemy came close to him, and in retiring kept firing

By that time, of course, it was presumed that the out-lying picquet, having summoned the in-lying picquet to its aid and warned the squadron, brigade and army of the attack, would have advanced to the support of the vedette, or at least made preparations to repulse the enemy as the vedettes withdrew. Practical problems sometimes arose, especially at night when a complicated series of passwords had to be established between the patrols and vedettes to prevent enemy infiltration. Sir Stapleton Cotton was in fact shot and wounded by a light cavalry vedette in July 1812 when he did not know the correct password, but overall the outpost duty was well performed, at least by the better units. As Sir Charles Oman points out in *Wellington's Army*:

> "much of the work of this kind speaks for itself. The most admirable achievement during the war was that of the 1st Hussars K.G.L., who, assisted by the 14th and 16th Light Dragoons, kept for four months (March – May 1810) the line of the Agueda and Azava, 40 miles long, against a fourfold strength of French cavalry without once letting a hostile reconnaissance through, losing a picquet or even a vedette, or sending a piece of false information to General Craufurd, whose front they covered".

Attacks did not always materialise, however, and in the pauses between operations the probing patrols were expected to engage in skirmishing with their enemy counterparts, both to prevent the collection of information or forage and to seek out possible weak spots in the French defences. As with reconnaissance and outpost work, the majority of light cavalry regiments proved to be well suited to this duty. Indeed, with the exception of slight problems over the accuracy of carbine fire when the enemy was engaged in a skirmish – a sergeant of the 7th Hussars told William Verner that on one occasion all the members of his patrol were "firing at an Officer who was mounted on a grey horse, but none could hit him" – there appear to have been very few difficulties over adaptability to such tactics. This was logical, for the light cavalry had been designed specifically for detached duty, and although this had been overlaid by the emphasis upon shock action, the basic features, recognised and dwelt upon by Hawley, Hinde and Simes in the eighteenth century, continued to exist. Once the initial difficulties which resulted from a lack of training had been overcome, therefore, the lessons of reconnaissance, skirmishing and outpost duty were relatively easy to learn and put into effect, particularly in the better regiments. Unfortunately, once again, these lessons were not applied to peace time soldiering, for as soon as the French Wars ended in 1815, they were quietly forgotten and the importance of the charge re-asserted. As Tomkinson sadly noted in 1819, "on our return to English duty we continued the old system, each regiment estimating its merit by the celerity of movement. I do not think one idea has been suggested since our return from service by the experience we there gained, and in five years we shall have all to commence again on going abroad".

Conclusion

Tomkinson's retrospective comments stand as a poor reflection upon the long-term value of light cavalry experiences between 1808 and 1815. When the Peninsula War began, the light dragoon and hussar regiments of the British army were ill-prepared for active campaigning, having lost their aptitude for the tactics most suited to their composition. They depended upon shock action to prove their worth and lacked training in the duties of pursuit, retreat, reconnaissance and advanced posts which they were called upon to carry out. Once the idea of the charge had been shown to be overambitious, these units were able to contribute at all to British victories, which speaks well for their adaptability and improvisation in the field. However this is more than cancelled out by the total disregard for these lessons once peace returned. The emphasis upon shock action, as Tomkinson points out, was reasserted with remarkable and almost incomprehensible speed after 1815, so that when the next European war broke out in 1854, the light cavalry was once again ill-prepared for action. The costly and foolish Charge of the Light Brigade at Balaclava on 25 October 1854 reflects this dangerous lack of preparation.

But this does not mean that light cavalry tactics during the Peninsula and Waterloo campaigns had no short-term value, for, so far as Wellington was concerned, the deficiencies of the regiments were recognised and their roles re-structured accordingly. Writing in 1826, the Duke stated that he had found light cavalry useful:

"first .... upon advanced guards, flanks, etc. as the quickest movers and to enable me to know and see as much as possible in the shortest space of time; secondly, to use them in small bodies to attack small bodies of the enemy's cavalry". Nevertheless, because they "would gallop (and) could not preserve their order", he found them "so inferior .... to the French .... (that) although I consider one squadron a match for two French squadrons .... I should not have liked to see four British squadrons opposed to four French squadrons; and as numbers increased, and order became more necessary, I was more unwilling to risk our cavalry without having a greater superiority of numbers".

The degree of light cavalry success between 1808 and 1815 could not be summed up better.

Bibliography

Maj. Lord Carnock (ed), "Cavalry in the Corunna Campaign: Extracts from the Diary of the Adjutant of the 15th Hussars", *Journal of the Society for Army Historical Research*, Special Publication No. 4, (1936).

L. Cooper *British Regular Cavalry, 1644-1914* (1965).

Maj.-Gen. D. Dundas, *Rules and Regulations for the Cavalry* (1795).

M. Glover, *Wellington as Military Commander* (1968).

R. Glover, *Peninsular Preparation* (1963).

Capt. R. Hinde, *The Discipline of Light Horse* (1777).

Light Dragoons, 1813-1815

Corporal in service dress 1813-1815
11th Regt.

Trumpeter in service dress 1812-1815
23rd Regt.

Sergeant in marching order 1813-1815
14th Regt.

Troop-Sergeant-Major
January 1815
11th Regt.

Sergeant
1815
12th Regt.

Trumpet-Major
1815
13th Regt.

Trooper in service dress 1815
13th Regt.

Officer	General-Officer of Hussars	Officer
1815	full dress	1813-1815
16th Regt.	1813-1815	23rd Regt.

Officer in review order 1812-1815
9th Regt.

Officer in marching order 1815
14th Regt.

Officer in service dress 1813-1814
12th Regt.

Officer in service dress 1815
13th Regt.

Maj. J. G. Le Marchant, *Rules and Regulations for the Sword Exercises of the Cavalry* (1796).

W. Müller, *The Elements of the Science of War* (3 vols, 1811).

Sir C. Oman, *Wellington's Army* (1912).

Maj.-Gen. Baron De Rottenberg, *Regulations for the Exercise of Riflemen and Light Infantry* (1798).

T. Simes, *The Military Guide for Young Officers* (2 vols, 1776).

Lt.-Col. W. Tomkinson, *The Diary of a Cavalry Officer, 1809-1815* (1894, reprinted 1971).

R. W. Verner (ed), "Reminiscences of William Verner (1782-1871) 7th Hussars", *Journal of the Society for Army Historical Research*, Special Publication No. 8, (1965).

J. Weller, *Wellington in the Peninsula, 1808-1814* (1962).

British Hussar